To Sylvia Kast: mother, wife, and breast cancer survivor
—D.J.K. & W.M.K.

To my mother, aunts, uncles, and cousins who are cancer survivors, and my family of supporters around them —DJ.F.

For my mom, who is the bravest and strongest woman I know —M.L.A.

A tribute to all the people around the world affected by cancer

Room to Read would like to thank Tatcha™ for their generous support of the STEAM-Powered Careers collection.

Copyright 2022 Room to Read. All rights reserved.

Written by Dr. Dieuwertje "DJ" Kast and Dr. W. Martin Kast
Featured scientist: DJ Fernandez
Illustrated by Michelle Laurentia Agatha
Edited by Jamie Leigh Real
Photo research by Kris Durán
Series art direction and design by Christy Hale
Series edited by Carol Burrell, Jamie Leigh Real, Jocelyn Argueta, and Deborah Davis
Copyedited by: Debra Deford-Minerva and Danielle Sunshine

ISBN 978-1-63845-060-3

Manufactured in Canada.

10 9 8 7 6 5 4 3 2

Room to Read
465 California Street #1000
San Francisco, California 94104
roomtoread.org

World change starts with educated children.©

STEAM-Powered Careers

ONCOLOGY

by **Dr. Dieuwertje "DJ" Kast** and **Dr. W. Martin Kast**

featured Scientist: **DJ Fernandez**

illustrated by **Michelle Laurentia Agatha**

Room to Read

Contents

Explore Oncology with Jae and Felicia	6
What Is Oncology?	22
Meet the Scientist	24
Learn More about Oncology	30
Word List	34

"I feel like science is everywhere," Felicia says to her friend Jae, flicking her tail.

"Why do you say that?" Jae asks.

"There are so many types of science out there. There is probably a science field for everything."

"I do feel like that's true," Jae says.

"I wonder if cancer science has its own scientific name too," says Felicia.

Jae glances at her. "Why do you ask about cancer science?"

"One of my human friends has cancer," Felicia says, "but I don't really know what that means."

Jae scratches Felicia's head. "I'm sorry to hear about your friend."

"All I know is that they're sick," she says. "I just want to make sure they're OK."

"I understand. Fortunately, many cancers can be cured," Jae says. "Right now, cancer scientists like my grandpa are working on finding even more cures to help people." Smiling, he adds, "I want to help people when I grow up too. Let's head into the clubhouse and we can talk about it."

People who get cancer are usually older. Cancers are rare in kids.

Oncology 9

"My grandpa has been teaching me a lot about cancer," he says. "It's an interesting science, and there's so much to study."

Felicia's eyes widen. "Can you tell me more?"

"Sure! Did you know there are over one hundred types of cancer?"

"Whoa!" says Felicia. "That's a lot."

"It is. Also, different types of cancers appear differently in the body. Some cancers appear as hard bumps and others can show on your skin and change in color and size."

Cancer types are usually named for the organs or tissues where the cancer grows. Brain cancer is in the brain and lung cancer is in the lungs.

"So, what exactly is happening in my friend's body?" Felicia asks. Felicia cozies up in front of the chalkboard and watches her friend draw healthy **cells** and cancer cells.

A tumor is a group of fast-growing cells. It can either be **benign** (not harmful, and doesn't move around in the body) or **malignant** (harmful, and moves around in the body).

"Our body is made up of cells," Jae says. "But when cells in the body grow too fast, they start to get too crowded. They can turn into cancer cells." He circles a drawing of a lot of cells.

Felicia nods, then points to a corner of the chalkboard. "What are these mean-looking cells over here?"

"Cancer scientists have found that cancer cells are like bullies in the body. These are the bully cancer cells that are growing too fast! They fight for space and pile on top of one another. This pile of cancer cells right here is called a **tumor**."

"Do these tumors stay in one place, or do they move around?" Felicia asks.

"They sometimes move around a person's body through their blood. I can show you! Let's go on an adventure inside the body!"

"All right, an adventure!" Felicia says, wagging her tail.

They put on their goggles and jump right into the chalkboard!
SWOOSH!

"It sure is sticky and red in here!" Felicia says, stepping around carefully. "What's that over there? Are those the bully cells?" She adjusts her goggles and takes a closer look.

"Yes, that's the tumor!" Jae says. "This one is particularly mean and wants to take up even more space in the body! On the skin, it looks like a bump."

"How do we stop tumors from taking up too much space?" Felicia asks.

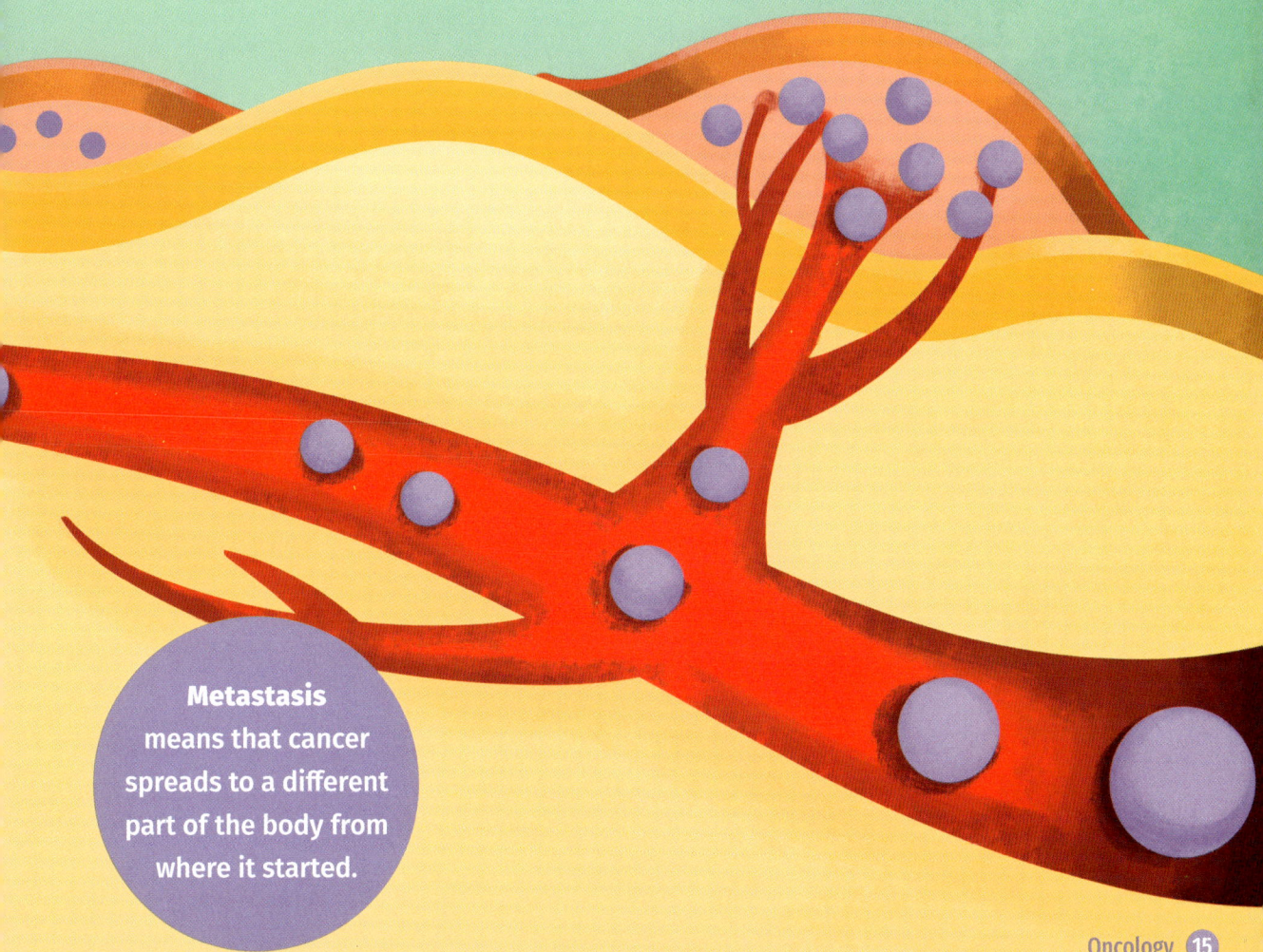

Metastasis means that cancer spreads to a different part of the body from where it started.

"Well, there are three main ways that doctors can treat cancer," Jae says. "Doctors can cut out the cancer during surgery, use medicine such as **chemotherapy**, or treat with **radiation**. Any of those treatments can kick those bully cancer cells to the curb."

"Interesting," Felicia says.

"Here's a cancer treatment room," Jae says. "It's important to wear very clean clothes in this room so we don't bring in germs that could make people sicker."

Jae washes his hands and puts on gloves, and they both put on surgical masks.

Felicia laughs. "I've never seen you so clean before!"

Jae laughs too. "You're right! Now look around. This is where the radiation treatment takes place."

"Wow! That's a big machine," Felicia says. "This is so cool! But what if my friend has radiation treatment and it doesn't work?"

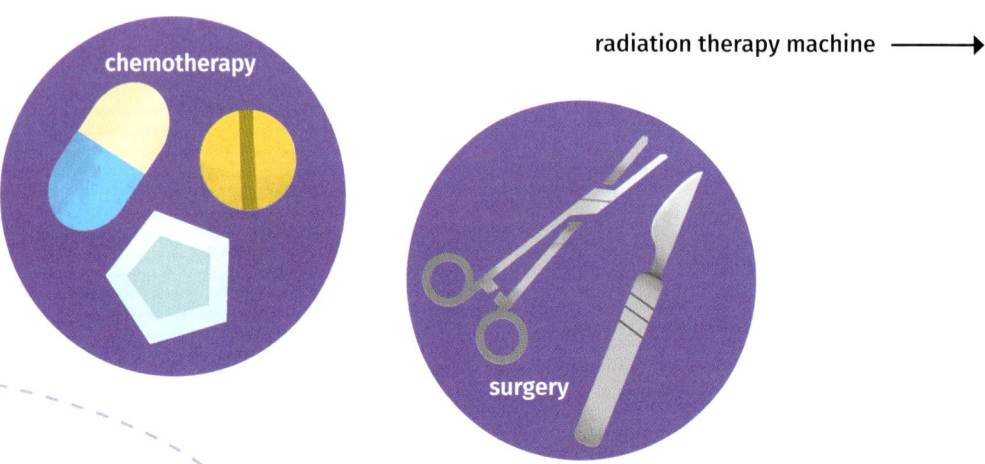

radiation therapy machine ⟶

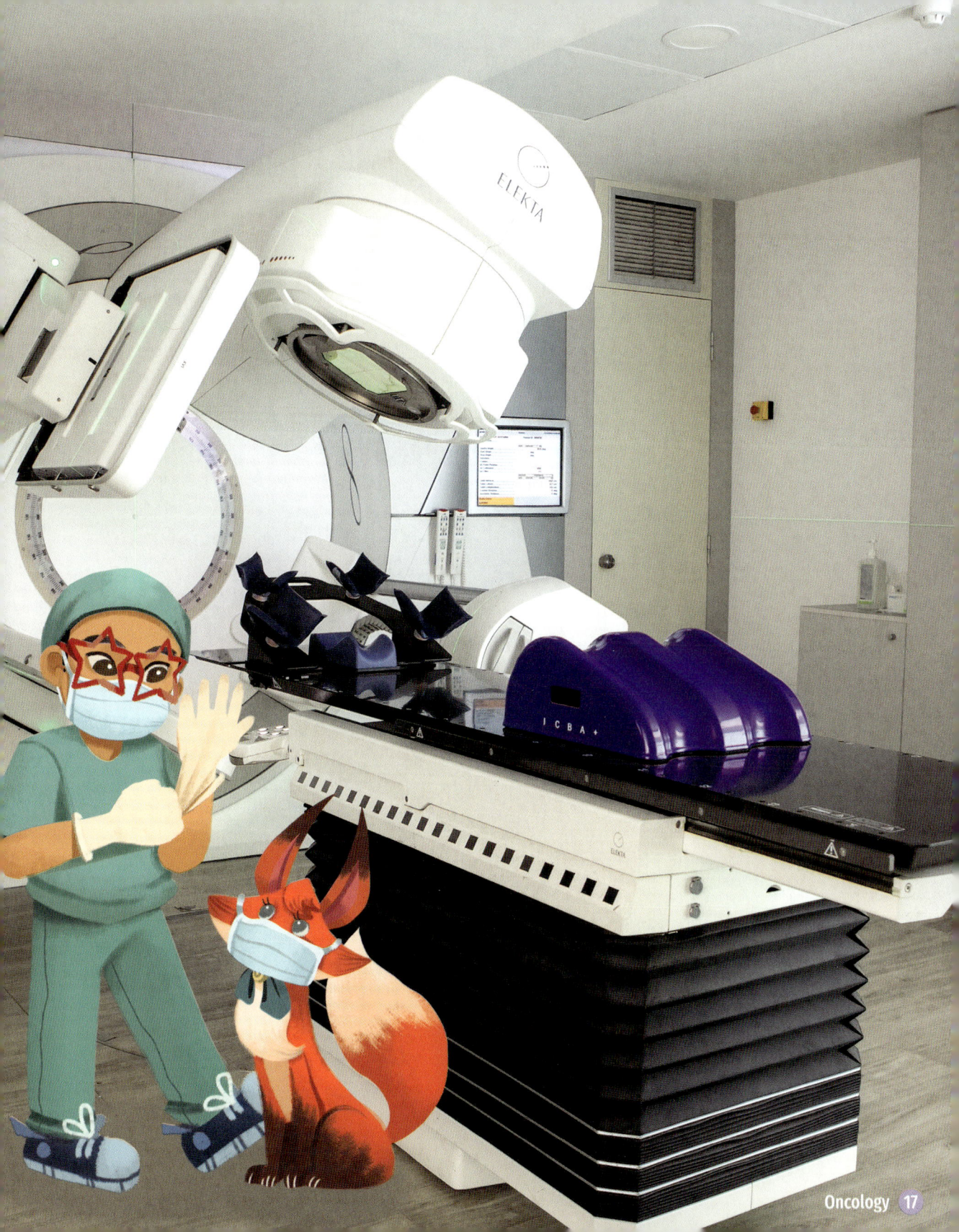

Oncology 17

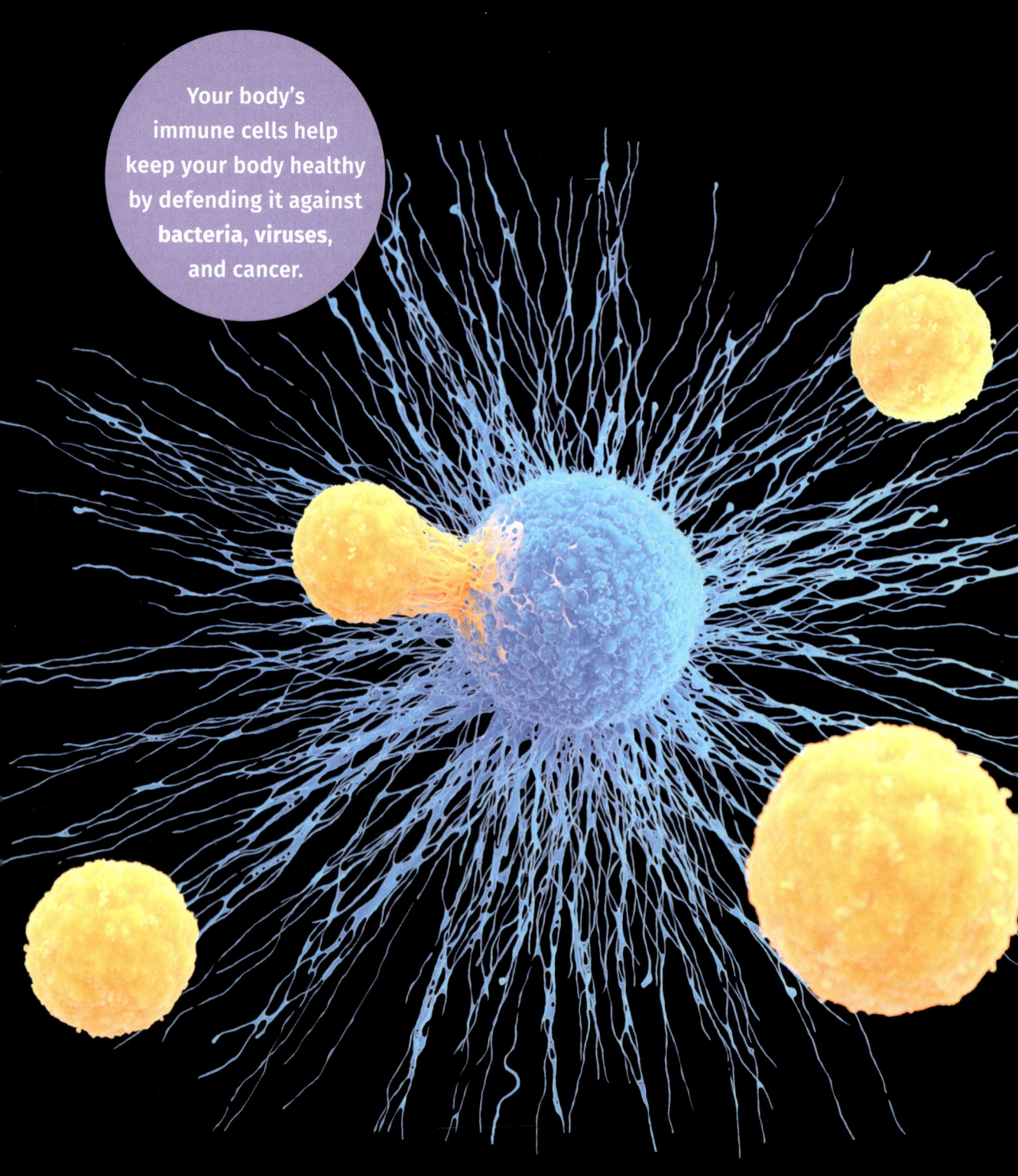

T-cell attaching to cancer cell

"That's a good question," Jae says. "Cancer scientists are researching new treatments, such as **immunotherapy**, that teach the body's **immune cells** to fight against these bully cancer cells. The two types of immune cells, called **B-cells** and **T-cells**, work together to kill only the cancer cells. They leave the healthy cells alone."

"I had no idea that there were tumors inside my friend's body, or that there were so many ways to fight cancer," Felicia says. "This makes a lot more sense now. But . . ."

Jae kneels next to his friend. She looks a bit worried.

"Can my friend give me their cancer cells?" she asks.

"Nope!" Jae says. "You can't catch cancer. It's not like getting a cold. You can hug your friend and let them know you are there for them."

Felicia looks relieved, and she smiles her little fox smile. "One more question before we head back to the clubhouse. Is there anything I can do to keep myself from getting cancer?"

"Well," Jae says, "there are some things you can do to reduce the chances of getting cancer, like eating healthy and nutritious foods, exercising regularly, wearing sunblock, and not smoking . . . There are even some cancers that you can prevent by getting a vaccine!"

"Got it," Felicia says. She wags her tail back and forth. "I'm happy your grandpa is doing this type of work."

"Yeah, it's pretty cool," Jae says. "He's helping a lot of people like your friend get better."

 STEAM-Powered Careers

Cancer scientists have found that people sometimes get cancer because of genetics, which means it runs in the family. Other times, it can be caused by living where the air or water is dirty.

Jae and Felicia jump back into the clubhouse with a big WHOMP! They are just in time for snacks! Jae scratches Felicia on her head again as they settle into the beanbags to munch on fruit.

What Is Oncology?

Jae and Felicia took us on a whirlwind tour of oncology. Before we meet **DJ Fernandez**, a scientist like Jae's grandfather, let's go over some terms that will be helpful in the lab.

Oncology (pronounced on-COLL-o-jee) is the study of cancer. Cancer refers to a number of different diseases.

> Science is full of long words based on Greek and Latin. *Onco* is Greek for "tumor" or "mass," and *tumor* is Latin for "to swell."

Oncologists are doctors who discover, identify, and try to cure cancer in patients. When an oncologist discovers and identifies the cancer in a person's body, this is called a diagnosis.

Cancer researchers are scientists who study cancer to come up with new medicines or treatments.

DJ Fernandez is a cancer researcher who is studying the human papillomavirus (HPV), a virus that causes several types of cancer and can be prevented with a vaccine. Let's ask him some questions, and then he'll show us around the lab!

Virus (green), cancer cells (red and blue)

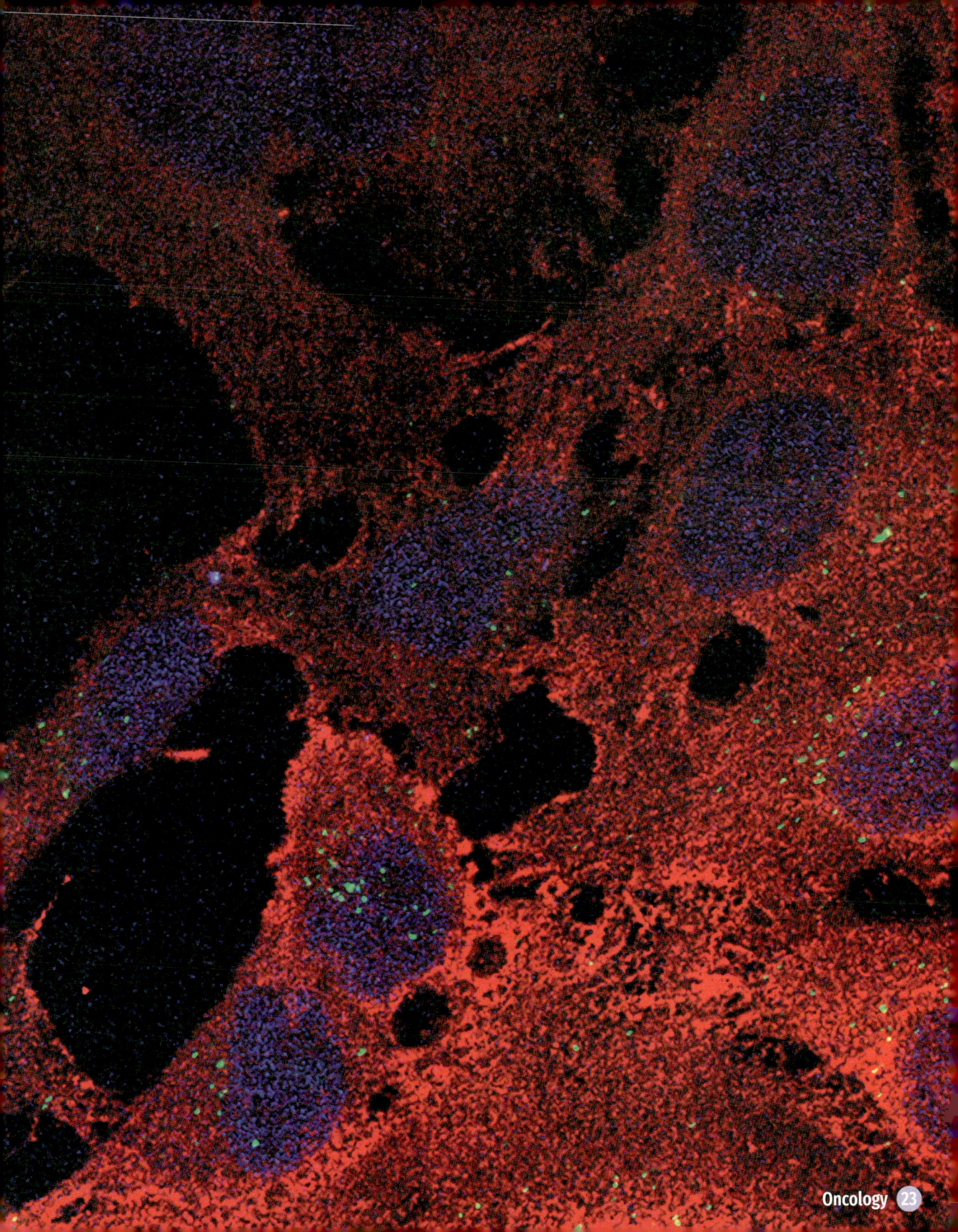

24 STEAM-Powered Careers

Meet the Scientist
DJ Fernandez

I love science because it is a way of explaining the natural world all around us. In college, I studied immunology & infectious diseases and toxicology (the study of things that are poisonous and how they affect people). I am currently a PhD student in medical biology.

Fun Fact #1: My mom is from Colombia, and my dad is from Honduras. My dad wanted me to be a lawyer because I asked so many questions as a little kid. For the same reason, my mom wanted me to be a scientist.

Fun Fact #2: I'm a DJ. I love making electronic music in my spare time and playing music for people! I go by the DJ name Crocodile Logic, and my music is inspired by biology.

What is your favorite thing about oncology?

What is your least favorite thing about oncology?

My favorite thing is helping people by looking for cures for cancer.

I wish it didn't take such a long time to find cures.

Let's check out my typical day . . .

A Day in the Life, Part 1

Every morning, I crawl out of bed and head downstairs to my home gym. After a good workout, I make a healthy and colorful breakfast (tomatoes, kale, and eggs).

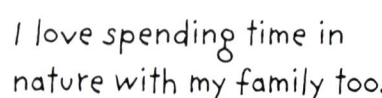

MEOW! MEOW! Prrrrrrrrrrrrrrrrrrrrrrrrrrrrrrrrrrr

I love my cat, Me-Mow, and I make sure to spend time in the morning with her before heading to the laboratory.

I love spending time in nature with my family too.

Me and Dad

26 STEAM-Powered Careers

I scoot to work on my Vespa. During ride, I think about my day ahead in the laboratory. I'm always excited find out the results of my experiments!

ZOOM!

NCCC

I am a scientist at the Norris Comprehensive Cancer Center of the University of Southern California.

There's a lot of work to do in the Kast Laboratory, where I study the human papillomavirus (HPV), a virus that causes several types of cancer. People can prevent these cancers by getting an HPV vaccine.

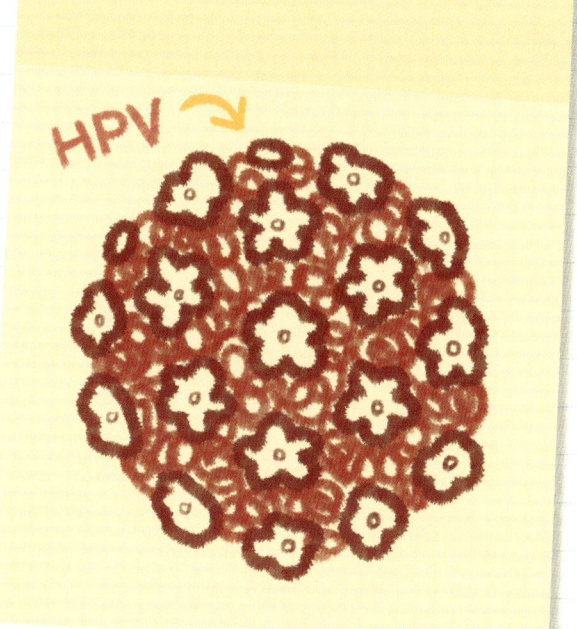
HPV

HPV

Oncology

A Day in the Life, Part 2

At the lab, I wash my hands before putting on my lab coat and gloves. The gloves protect me and keep me from contaminating my experiments.

I'm curious about how HPV enters the cells that it infects. I study the parts of a cell that HPV uses to get in. I watch HPV enter cells and try to block it in different ways. To do this, I use special microscopes.

I'm working with **DNA** in this picture. I have to keep the work in an enclosed cabinet like this one to make sure the experiments don't get contaminated with my own DNA.

STEAM-Powered Careers

This red liquid contains all the nutrients that cells need to survive. When you work with cells and viruses, you have to be very careful and focused. I try to do my best when doing experiments so that my discoveries can help cure cancer.

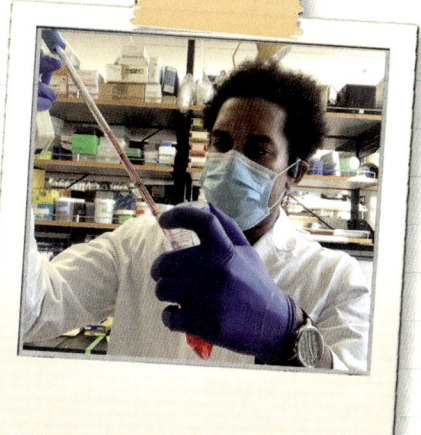

Any results yet?

Dr. Kast likes to see what I am working on in the lab, especially when I discover important findings!

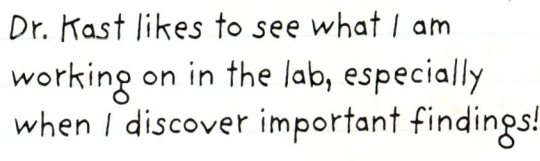

We are very close to finding a new piece of the puzzle!

I think viruses are extremely interesting. Viruses like to infect us, so we study them in the lab to learn how they do this. At the same time, we learn about our own bodies.

My favorite thing about research is that each day is a new chance for discovery. When I make a new discovery, I get to see something that no one in the world has ever seen before.

Oncology 29

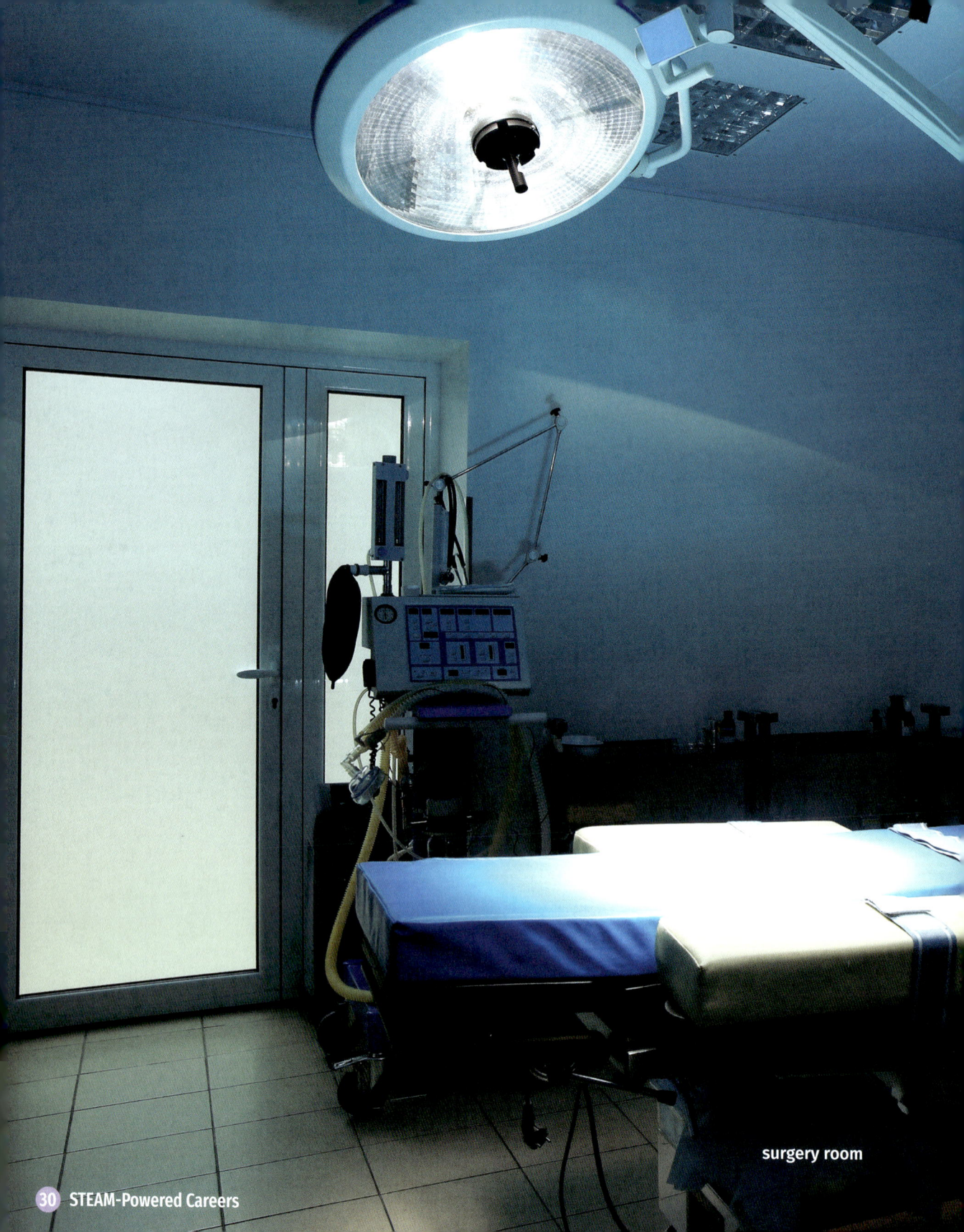
surgery room

STEAM Careers in Oncology

All these fields of study help us understand and treat cancer!

There are many fields to choose from if you want to work in oncology. Anatomy is the study of a body's structure, and biochemistry is the study of molecules (the parts that make up cells) found in living organisms.

Cell biology is the study of cells. Since cancer is a cellular disease, the search for cures often happens here too.

Medicine is a large field that focuses on healing and well-being. Pharmacology is the study of how medicines affect your body. Health and disease prevention, or public health, focuses on keeping people healthy and stopping disease from spreading.

With these jobs you're helping people live longer and healthier lives and working towards finding a cure for cancer.

The Future of Oncology

Oncologists and cancer researchers work hard to make cancer a disease of the past. They study how cancer is different based on who you are and where you live and make medicine that is designed especially for you and your body. They also find ways to teach your immune system what it needs to do to get rid of cancer.

As they discover more ways to find cancers early, cancers will have less time to spread. Computers can also help them understand how cancers spread and how to stop them. Oncologists and cancer researchers also work on vaccines to make sure that some cancers don't grow at all.

objective lenses and stage of a microscope

STEAM-Powered Careers

Do You Want to Be an Oncologist?

Learning more about the human body and how it works is a great place to start. Your own doctor might even be able to answer some questions or connect you with someone who can. In the meantime, you can try your best in school, take math and science classes, and explore the world around you.

And you can find out more about related jobs such as:

- professor or researcher at a university
- doctor or physician's assistant (PA)
- nurse
- biotechnology company researcher
- pharmaceutical company researcher (for a company that creates medicines)
- lab technician
- science communicator (someone who writes articles and books or who does presentations or makes videos to tell the public about the latest science information and news)

Word List

benign: not harmful

B-cell: a type of blood *cell* which produces antibodies that fight bacteria, viruses, and cancer

cell: a small building block that makes up all living things

chemotherapy: medicine used to treat cancer

immune cell: a *cell* that is part of the immune system, which fights diseases and infections

DNA: the material in living things that is passed down from parent to child and carries instructions for development, survival, and reproduction

immunotherapy: a type of prevention or treatment of disease that teaches the body's *immune cells* to fight cancer *cells*

malignant: a scientific word meaning harmful

metastasis: a scientific word that describes how a disease, such as cancer, can spread throughout the body

radiation: a form of cancer treatment that uses rays of energy

T-cell: a type of *immune cell* that kills cancer *cells*

tumor: a group of fast-growing *cells*

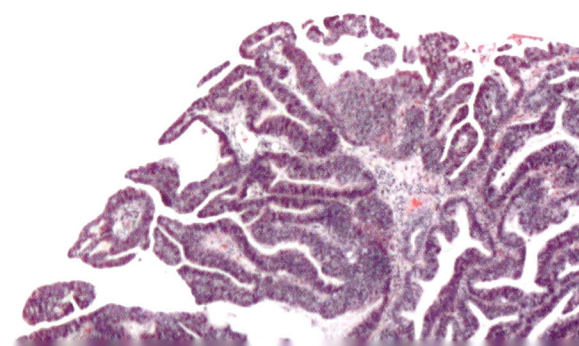

Oncology Resources

Check out these books:
Nowhere Hair by Sue Glader

What Is Cancer? A Book for Kids by Carolina Schmidt

Skin Cancer by Marjorie L. Buckmaster

Helpful website to understand cancer:
"What Is Cancer?" *KidsHealth*
    https://kidshealth.org/en/kids/cancer.html

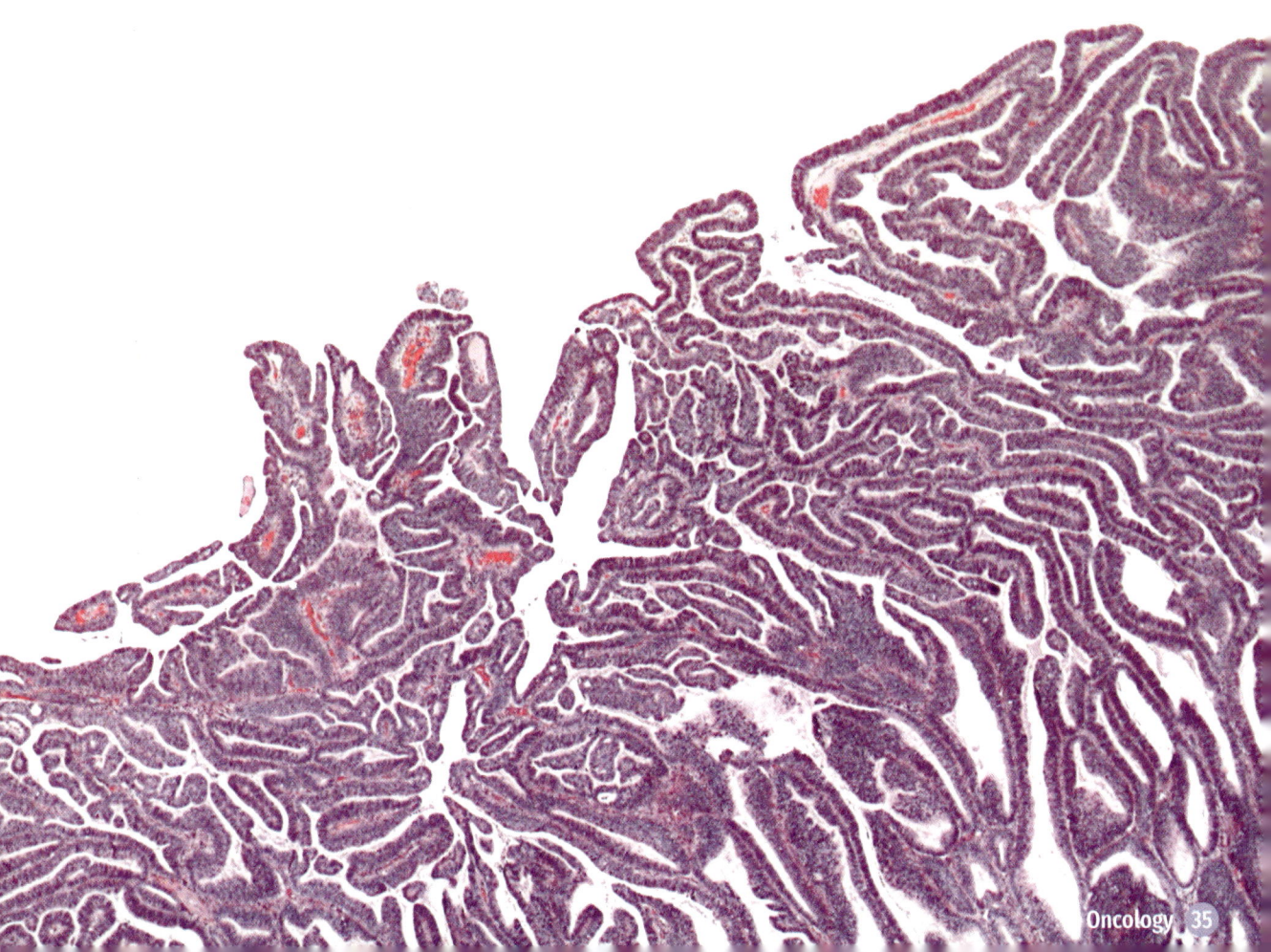

Acknowledgments

University of Southern California

Dornsife College of Letters, Arts, and Sciences; Joint Educational Project STEM Education Programs and Keck School of Medicine; Norris Comprehensive Cancer Center

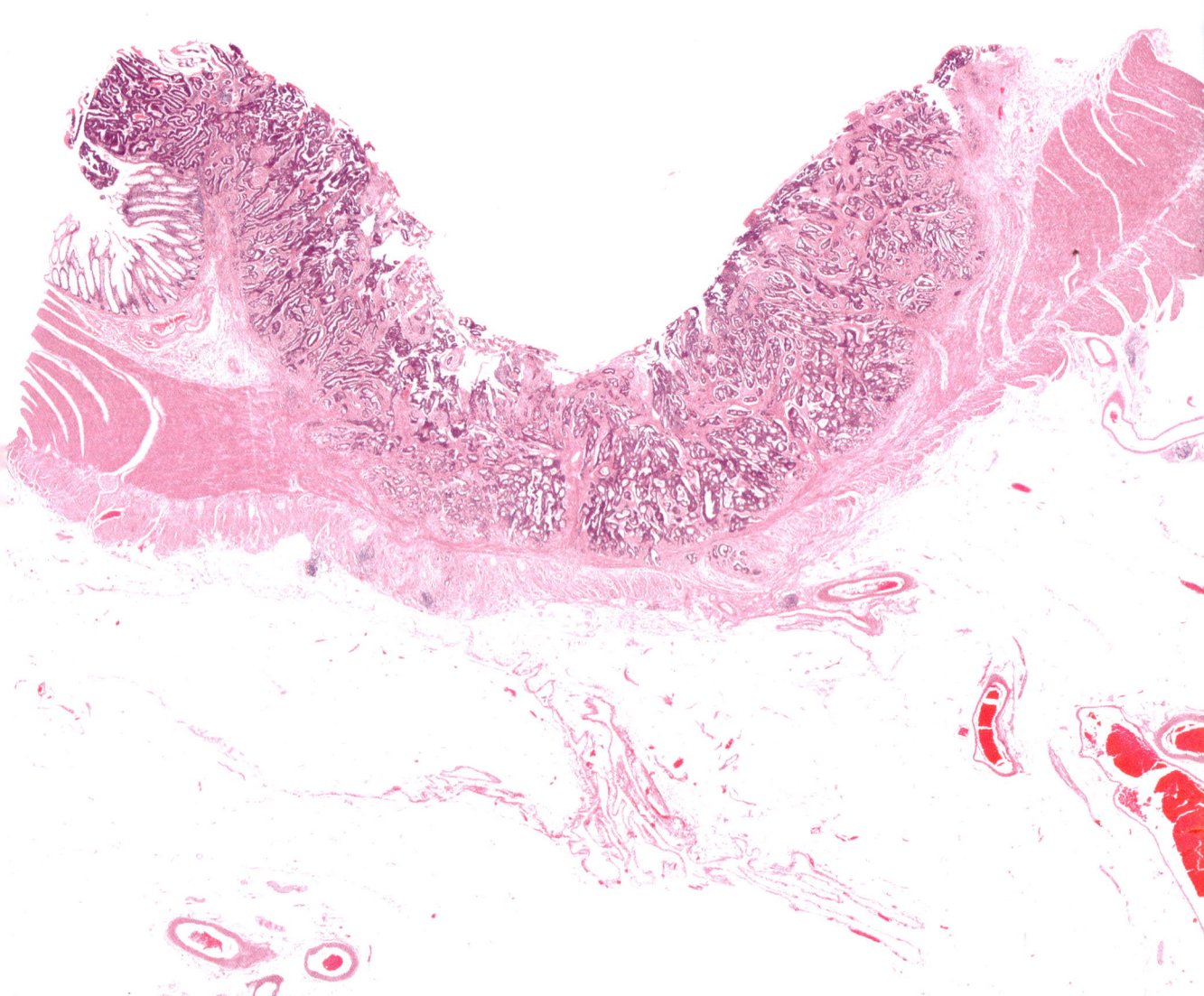

Dr. Dieuwertje "DJ" Kast, EdD, is the director of STEM Education for the USC Joint Educational Project, based in Los Angeles, California. She holds a doctorate in education from USC, where she focused on teacher education in multicultural societies in STEM. Her mission is to level the playing field for underserved students in STEM.

Dr. W. Martin Kast, PhD, is a cancer researcher at the USC/Norris Comprehensive Cancer Center in Los Angeles, California. He teaches the science behind cancer to medical, graduate, and undergraduate students, as well as high school, middle school, and elementary school students.

DJ Fernandez, a PhD student, sees the world through a scientific lens and applies the scientific method to his daily life. When he's not conducting research, teaching science, or advising, he's on a hike or at the beach, making observations of the natural world for fun. He plans on doing research until he grows old, and continuing even then.

Michelle Laurentia Agatha was born in Jakarta, Indonesia. Ever since she was young, she has had a huge interest in cartoons and illustrated books. Michelle pursued her dream of becoming an illustrator by earning a Bachelor of Fine Arts degree from the Academy of Art University in San Francisco. Currently, Michelle is working as a children's book illustrator, concept artist, and UI/UX designer.

Explore the Complete

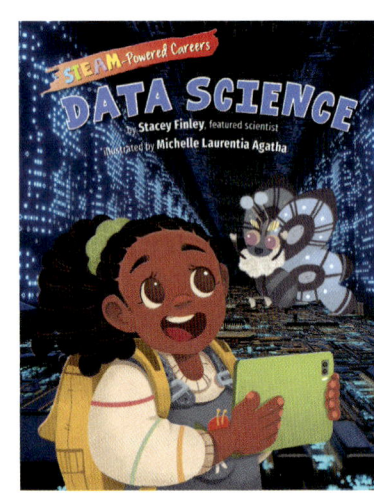
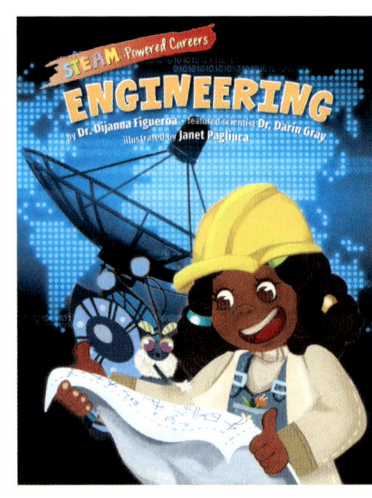

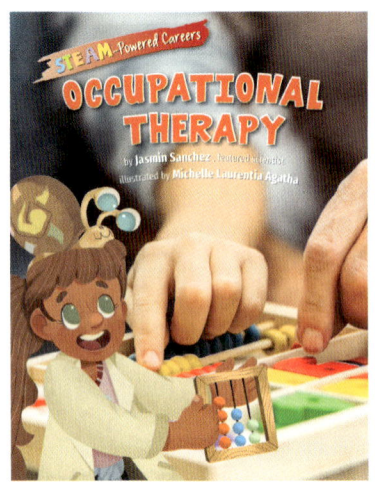
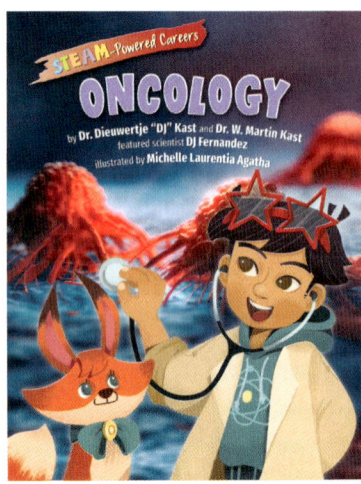

STEAM-Powered Careers Series!

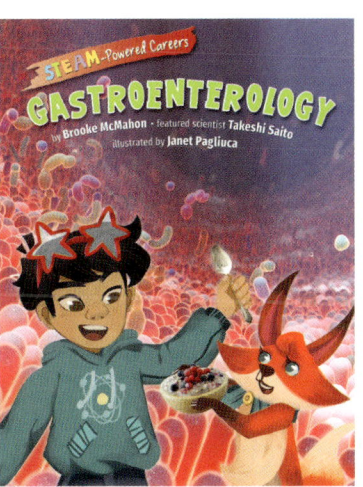
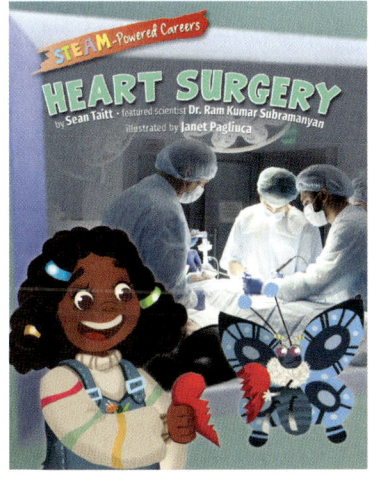
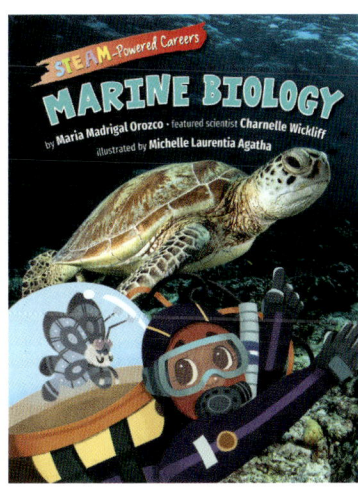
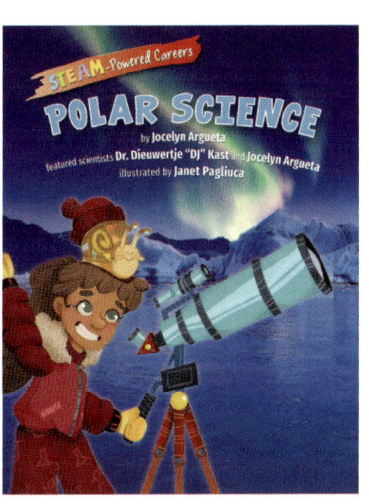
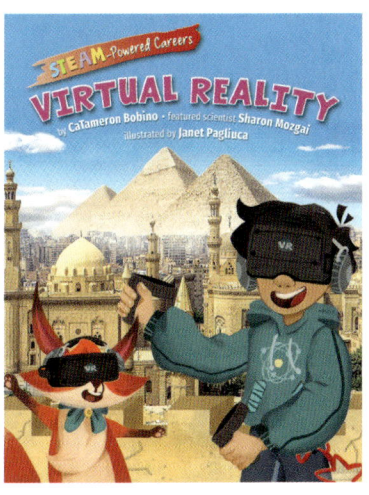

Photo credits

Cover iStock.com/koto_feja **16–17** vverve/Depositphotos.com **18** Roger Harris/Science Photo Library **20** ivoxis/Pixabay; Aptyp_koK/Depositphotos.com **22–23** photo courtesy of DJ Fernandez **24–25** Kim Lühen; photos courtesy of DJ Fernandez **26–27** Clayton Rosa; photos courtesy of DJ Fernandez; Kim Lühen **28–29** Werneuchen, public domain, via Wikimedia Commons; Kim Lühen; photo courtesy of DJ Fernandez **30–31** NewAfrica/Depositphotos.com **32–33** CC0 public domain, via Pxhere **34–35** Nephron, CC BY-SA 3.0, via Wikimedia Commons **36** "Colon Adenocarcinoma, Whole-Mount Scan," Ed Uthman, CC BY 2.0, via Flickr **37** photo courtesy of Dr. Dieuwertje "DJ" Kast; photo courtesy of Dr. W. Martin Kast; photo courtesy of DJ Fernandez; photo courtesy of Michelle Agatha **40** photo by Anna Shvets from Pexels